This Is a Let's-Read-and-Find-Out Science Book®

Glaciers

Revised Edition

by Wendell V. Tangborn
illustrated by Marc Simont

Thomas Y. Crowell   New York

Other Recent Let's-Read-and-Find-Out Science Books® You Will Enjoy

Dinosaur Bones • Snakes Are Hunters • Evolution • Danger—Icebergs! • Rockets and Satellites • The Planets in Our Solar System • The Moon Seems to Change • Ant Cities • Get Ready for Robots! • Gravity Is a Mystery • Snow Is Falling • Journey into a Black Hole • What Makes Day and Night • Air Is All Around You • Turtle Talk • What the Moon Is Like • Hurricane Watch • Sunshine Makes the Seasons • My Visit to the Dinosaurs • The BASIC Book • Bits and Bytes • Germs Make Me Sick! • Flash, Crash, Rumble, and Roll • Volcanoes • Dinosaurs Are Different • What Happens to a Hamburger • Meet the Computer • How to Talk to Your Computer • Comets • Rock Collecting • Is There Life in Outer Space? • All Kinds of Feet • Flying Giants of Long Ago • Rain and Hail • Why I Cough, Sneeze, Shiver, Hiccup, & Yawn • You Can't Make a Move Without Your Muscles • The Sky Is Full of Stars • Digging Up Dinosaurs • No Measles, No Mumps for Me • When Birds Change Their Feathers • Birds Are Flying • Cactus in the Desert • Me and My Family Tree • Shells Are Skeletons • Caves • Wild and Woolly Mammoths • Corals • Energy from the Sun • Corn Is Maize

The *Let's-Read-and-Find-Out Science Book* series was originated by Dr. Franklyn M. Branley, Astronomer Emeritus and former Chairman of the American Museum–Hayden Planetarium, and was formerly co-edited by him and Dr. Roma Gans, Professor Emeritus of Childhood Education, Teachers College, Columbia University. For a complete catalog of Let's-Read-and-Find-Out Science Books, write to Thomas Y. Crowell Junior Books, Harper & Row, Publishers, Inc., 10 East 53rd Street, New York, NY 10022.

Let's-Read-and-Find-Out Science Book is a registered trademark of Harper & Row, Publishers, Inc.

Library of Congress Cataloging-in-Publication Data

Tangborn, Wendell V.
  Glaciers.

  (Let's-read-and-find-out science book)
  Summary: Explains how and where glaciers form, how they move, and how they shape the land.
  1. Glaciers—Juvenile literature. [1. Glaciers]
I. Simont, Marc, ill. II. Title. III. Series.
GB2403.8.T36 1988   551.3'12   87-47696
ISBN 0-690-04682-0
ISBN 0-690-04684-7 (lib. bdg.)

  (A Let's-read-and-find-out book)
  "A Harper Trophy book."
ISBN 0-06-445076-7 (pbk.)   87-45306

To John, Andrew, Inger and Eric—when they were children

Winters are very cold in many places. Hills and valleys are covered with snow. Big lakes freeze. Ice covers rivers from shore to shore. If the winter stays cold, the ice gets thicker and thicker. The snow piles deeper and deeper.

In springtime the snow melts from hills
and valleys. The ice disappears from lakes
and rivers.

But some places on the Earth are cold all the
time. There snow does not melt away even in
the summertime. On very high mountains, for
example, you can see snow and ice all through
the summer.

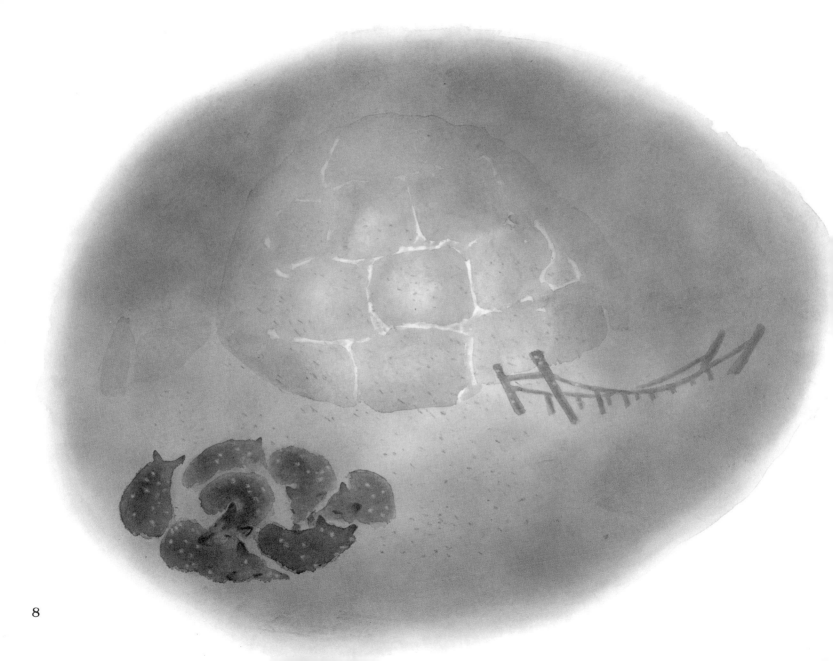

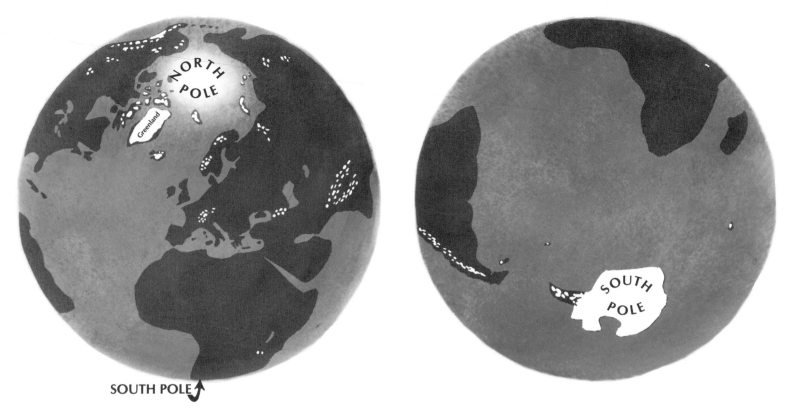

SOUTH POLE↑

Greenland, near the North Pole, is covered by ice all the time. At the South Pole miles and miles of thick ice cover the land. The snow and ice never melt away. Year after year more snow falls. More snow piles deeper and deeper. Snow that does not melt year after year turns into ice.

9

The ice piles up for hundreds, even thousands
of years. Year after year it becomes thicker
and thicker. The ice may be ten feet thick,
a hundred feet thick. In some places the ice
is a thousand feet thick, and sometimes many
thousands of feet thick.

Big fields of thick ice are called glaciers.

Glaciers move. The weight of the ice makes a glacier move. It moves down the mountains toward the sea.

Glaciers move so slowly that you cannot see they are moving. Some glaciers move ten feet in one day. Some move less than an inch a day.

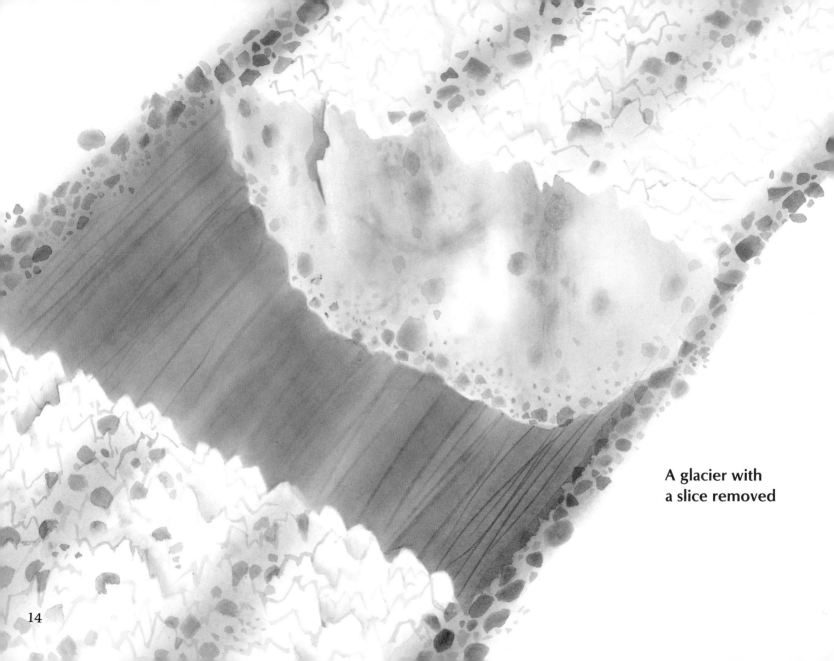

A glacier with
a slice removed

Nothing can stop these big fields of ice from moving. As they move over land they pick up stones and boulders. As they move through valleys they cut them deep and wide. The ice is packed with soil and trees, rocks and boulders. Rocks and boulders, soil and trees are ground together for hundreds and thousands of years. Some boulders are ground as fine as flour.

Slowly, slowly the glaciers move, grinding and crushing rocks, hillsides, trees, and forests. A glacier could push a whole city out of its path. Glaciers move on and on, year after year after year.

Sometimes the ice stretches and makes huge cracks. The cracks may be a hundred feet deep. Some are so wide you could not throw a stone across them.

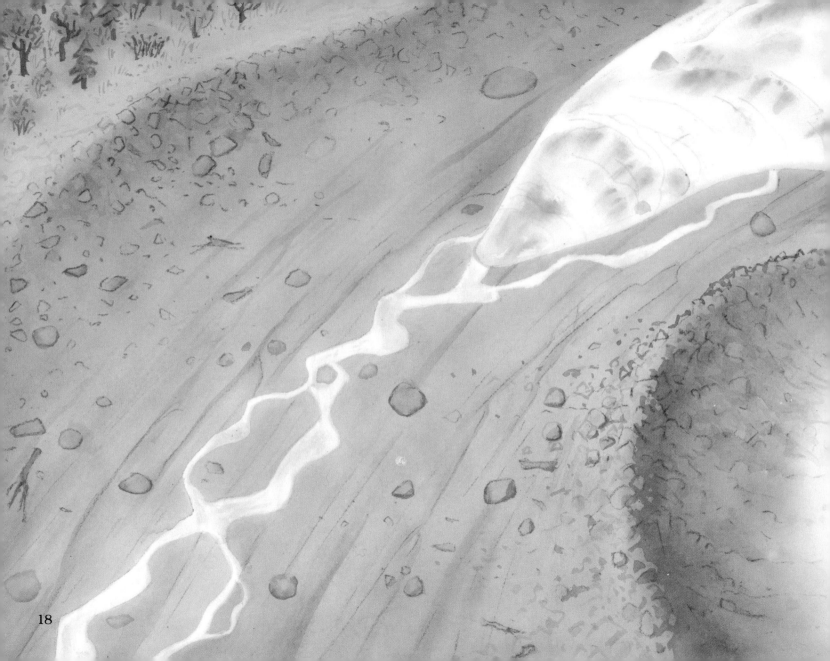

When the edge of the glacier comes to warmer weather it slowly melts. Streams run from the melting edge. The streams may be milky white from the ground-up rocks.

As the glacier melts, rocks and boulders, trees, and tons of soil are dropped. Rocks and boulders, trees and soil may be piled into long, low hills. These hills are called moraines.

Some glaciers that form in the far north and far south move into the sea. Gigantic pieces break off and become icebergs. These icebergs float out to sea.

When the icebergs melt they drop soil, sand, rocks, and boulders into the ocean.

Thousands of years ago glaciers covered large parts of the Earth. These glaciers were formed during the great ice ages. The white areas on the map show where the Earth was covered by glaciers during the ice ages.

Today glaciers are found only in the polar regions and on high mountains.

NORTH POLE

Maybe the place where you live used to be under a glacier. Rocks you pick up may have been dropped by a glacier as it melted. They may have been dragged from a faraway place. A big rounded boulder that stands all alone may have been carried by a glacier. Some of the hills you slide down may have been made by a glacier long, long ago in an early ice age.

Today parts of the Earth are covered with snow and ice all through the year. Many glaciers that we see are still being formed. The Earth may still be in a Little Ice Age that began about seven hundred years ago.

Glaciers are scraping up rocks and boulders, trees and soil, grinding, grinding, grinding. You can see the places on the map.

NORTH
POLE

Greenland

North America

South America

Asia

Europe

Africa

As you travel, look for places where there are glaciers today. Look for places where there were glaciers long ago. Look for pits where gravel is dug. The gravel was made by glaciers.

Look for hills like these. They are made of rocks and soil left behind when a glacier melted. Look for deep scratches in rocks. They were made when rocks in glaciers were scraped together. Look for huge boulders that stand all alone.

Much of the land on which we live was
shaped by glaciers long ago.